One of the key foci of mechanical statistics is to derive conclusions from what should be a large amount of "representative" data. This helps make acute claims about something. The accuracy of a poll. The likelihood of a natural disaster. And so on. Traditional statistics models further generally rely on specific types of distributions, which in essence serve -rightly or wrongly- as a rigid blueprint for a phenomenon.

While we often think we do, we many times don't have all the data we need. We may lack a significant portion of the data actually. Like spending your time admiring a stunning ocean but neglect to see a shark hiding underneath (or behind other data!). So while we may obtain results, one may also lack a real understanding of what is important. You might get false personal confidence from your own output; though you may never be able to replicate the said result.

Further, we are also obviously not all clones of one another. The implications of this implicitly run deep, even within the realm of math. We interpret similar data in different ways. We vastly disagree about what are the relevant facts, or what is the best analysis or approach. We ask different questions besides "is the glass half-full or half-empty?" These blind distinctions in our socio-economic backgrounds and knowledge also amplify to large quantitative differences in our opinions of the data itself.

The processes associated with traditional statistics can leave one with a protracted and confusing sense about how right we are in our interpretation of numeric ideas. For example, we can oft-times both misuse, and abuse statistical models, which ironically are the same ones we cherry-picked. Or ones that we self-decorated, and use only because they would show us the results we want. Analysts have had us fooled by their own models over a long history. In recent decades, we see examples with Enron and Madoff, WMDs, the World Trade Center, 737 Max, presidential polling, coronavirus, and many others.

In all these cases, the subjective modeling and trust in the analysis of someone in authority didn't always make things clearer. Rather this approach incorrectly reinforced flawed data, or the analyst's own analytical confusion. The six topical vignettes of this book center on missing data, proximate bias, perception wedges, busted rulers, compounded errors, and simulators.

We begin by laying the groundwork for not the data itself, but rather how different people are disposed to thinking about similar data differently (and why that's an key understanding as you engage with the world!) We explore what is meant to look at the quality of the underlying data, which you might want to analyze. We also talk about the open-mindedness of the analyst, and the legitimacy of either in analyzing a certain domain. This is something traditional coursework has left a void, in terms of training young students.

We also aimed for this book to be visually enticing, in order to aid your overall enjoyment and experience. There are no formulas. We hope that you find the book fun and accessible, including in its efficient communication of critical themes. Please recommend it to others who might benefit!

Our upbringing is an often-overlooked aspect of statistical analysis, because it relies on information that is not assumed within mathematical formulae. This then, over time, exhibits itself in the way many companies prioritize diversity and inclusion in their executive ranks. How does one interpret a viral video, say, of a white police officer arresting a minority? How does an AI software interpret this same information, if not also tainted by the background of the tech person programming that AI model? And wouldn't all consumers be wrong when we later "admit" the video was only a small section of the entire context of what initiated the police encounter? These are the layers of continued statistical clarifications that are needed to be considered. And they add relevant vagueness, to any initial interpretation, for someone seeing that viral video.

So when we consider issues in our lives, we should spend more time thinking about the type and relevancy of the data. Do we are really have much of it to begin with? It's key to know what critical data is missing and (unbeknownst to us) causing us confusion. We should be modest with our causal interpretation of statistical results as well, which may basically be completely improbable or conversely far more "probable" once you consider all of the presented data could be missing pieces of simply be faulty. Again, we have seen this in the 2016 presidential polling, and in the ongoing coronavirus case data. We see it when studying personal Lottery expenses, but don't know how many people don't play at all.

There is fascination and rigidity towards connecting random data points from the most readily available data. Then assuming that they – collectively- hold complete insight into the future. And later, precariously, we may leverage this myopic misunderstanding when insisting others must also agree with it. We see it with disasters, elections, etc. Perhaps you have heard expressions similar to these:

"We used to have pandemic type diseases more often prior to world war. Therefore we have generally made progress on stopping pandemics."

"We just had a new pandemic, the first in a century. The last one had multiple waves, so we need to be on the look out for a second larger wave. Or we don't need to worry about another pandemic for another century."

"South Korea handled their pandemic better than the United States, therefore South Korea is the new gold standard in pandemic response."

What do these previous expressions have in common? They are meaningless statistical statements with little or zero practical sample size. Also appreciate that pandemics are also not tracked uniformly across globe, across history, or both.

These issues are also true in financial markets. Say when studying volatility data, where the world's most popular index "only" goes back to 1990. Even though all of the relevant financial history is longer than any of our lives! For example, the violent Black Monday of 1987 is excluded from any risk series that starts in 1990. Let alone any large crisis from a century ago. Some asset classes, or definitions of emerging markets, have also been added and removed from the economy, over time. What non-trivial impact would that have on our understanding?

Now here's a theoretical scenario about streaks. Say someone at a casino rolls a die with a "2", two times. You go over and see this player who then rolls "2" for a third time. What is the probability of what you just witnessed?

Was it a 1/216 chance of occurrence? Was it a 1/6 chance… or even a 1/36 chance? Can we truly ever know?

What does it say, that you only interested in player after the streak of 2 "2"s? Or, that only starting calculate the odds after the 2s are broken? for example, became this initial straight you may be to streak streak of

Always be cautious of statistical results that hinge upon specific recent relationships. You may do this because at times it is the only available data, or maybe because of muscle memory (e.g., you might be enamored to have just flown past a fabled unicorn, though it might reflect little about other people's experiences with same). Always ask yourself, to what degree are events being seen a surprise. And if they are eternally independent events, or not. In probability theory this is sometimes loosely captured with the crude concept of degrees of freedom.

Analyzing data

by assuming simplified properties   Critical perils

Why does one bother modeling a presidential election, when there are a high number of undecideds, or late-deciders, or disenfranchised voters… and all this occurring for months prior to the election? Where the very precise signal is subdued by the noise.

How do you determine how to evaluate the prudence of bitcoin, when no one knows how to model its worth or its return, given the very high variance and unknown underlying mathematical distribution? You simply can't.

How can anyone provide a constructive probability of dying from coronavirus in 2020, without first assessing the avoidance risk of each family member in their home? While it was easy to see the turbulence from China in January 2020 sputtering out of control, why were U.S. policy makers so sure they had contained the virus? Similar to bitcoin, the difficulty in modeling for some is the appreciation that they are witnessing a multiplicative process (much larger and noisier than the typical exponential growth processes, which themselves are highly volatile growth rates).

How does an MD/PhD-type either throw up their hands, or explain complicated multiplicative failures (or a break of success) to a boss who also isn't trained in the same? Say, a health-care system that would be overwhelmed by compounded problems that feedback loop onto itself? How do you plan for future moral-hazard behavior or of side-effects (of any significance), similar to how the government was unable to predict the long-term safety of the air at the World Trade Center? In the section illustration on the previous page, how to you know the impact of blowing-up a single mine?

..such as a fuse hidden beside compounded errors.

The resulting moves are simply entropy at times. The shocking win and impeachment-level finger-pointing to many of a presidential election, the "shocking" turning point of bitcoin, the ongoing twists in death forecasts due to coronavirus (both here and around the world). There are no easy answers when a model would simply not suffice. We have the elegance of convolution theory for example (both parametric and nonparametric versions), and mixed-model theory (parametric formulae discussed in my previous book Statistics Topics), though again suffer from the difficulty from wanting to presume a precise causal, and decision making information when the data is simply sparse at best; missing at worse. You may ultimately be modeling noise.

Now as coronavirus cases have started to presumably "bend" down in large swaths of the world during summer 2020, we are equally puzzled by all the ways we obtained that improvement (or was this improvement during "wave 1" simply a statistical mirage?) Explanation shrouded in data quality issues, narrow-minded politics, selection-bias, and randomness. Were the lockdowns too rigorous or were they adequate? And crucially, who determines that?

How much of an impact was weather, masks, social-distancing, etc? Does it matter if the models are different from state to state, and country to country? Are the people there approaching the virus differently at times (e.g., their data and testing)? Are there subtle mutations that are continuing to happen? Are our cultural response to defeat the virus appropriate? All of these details matter for statistics, yet analysts choose to focus on a small set of formulas that mostly ignore it.

There is the slow and growing disconnect, between one's perception and reality, based on a singular or narrow set of facts or analysis. It is truly the weakest aspect of "mastering" statistics: clutching to specific ideas at the expense of broader non-statistical gyrations that are equally critical. The longer and slower this wedge grows, the more confident that people are that they are on the right side of their perception. Profitable Robinhood accounts thinking they are geniuses for calling one turn of the market, minorities thinking discrimination is the sole cause of all their performance upsets, etc. The longer it takes, the more people avoid conceding the eventual gap between their incorrect understanding, and reality. This also happens, well after events have clearly turned. The coyote who speeds across the desert so fast. And then not realizing -but rather conceding- it has sped well off the cliff.

Sometimes this is due to the dynamics not being quantitative or predictive enough, or pride over-riding the underlying facts. What's critical here is to appreciate that the initial setting or pattern can often have some aspect of truth to it. But also, our individual contribution to that successful or failure pattern might be nil (and yet we instead continue to strongly accept or reject that result based on pride in our initial assessment).

Let's say it's the first day at a new school, and you love it. The second day of a new school, everything is still fine. Now you have a new budding expectation of a pattern in how you like this new school is. But on the third day, say something suddenly goes awry. Instead of acknowledging (or knowing how to acknowledge) this wedge (gap between your perception and reality) you may make-up excuses for why the new school is still really great. Your heightened expectations temporarily override the new evidence. You may start changing the bar and not in a way that ultimately benefits you. You start making excuses such as "this is really what great is", or "this is normal and the later part of the day will make up for it".

Recall public facts that mirrored a pandemic by January 2020. Our website was among a few who assessed catastrophic risk then. But the China government, WHO, financial market participants, and Trump didn't concede anything close, until over a month later. After the public dragged out the truth. On an aside, also notice that pandemic risk is covered in across different themes of this book.

The magic of wedges is also for people who use their imagination to continue to behave irrationally for far too long, only to change when it is most disadvantageous to do so. Universities with above-average tuition, government with below average interest rates, irrationally increasing housing prices, etc. Imagine through life that disequilibriums can continue to unsustainably grow and stay unchecked. When emotions are involved, it can be difficult to work with the right statistical models.

Statistical deductions are only as good are the accuracy of your contextual reference. In this area of the book, we will discuss the importance of you being humble, and true to yourself.

We know many times information we use is not as it appears. And this can result in laypeople being forced to use statistics in life, only to repeatedly chase after false trends. We also know from social media, that "trends" there can be engineered either by users, or by technology companies.

Say for example people constantly (and incorrectly) shouting out that an avocado is red. It's not, but the same people are not demanding others to acknowledge it anyway. Soon enough people believe what is false, and then go on to start questioning those who ever claimed it was something else (e.g., its actual color of green).

Just as quick, others will use this new ruler of a "red" avocado, to judge other objects (alligators, trees, etc) as now being red.

Perhaps you can relate in your own life to an unfair boss changing his or her mind about the standards for promotion, with the unfair goal to skip over you. You were always the shining and desirable red apple, but not judged as one in the unfair system that badly made people out to believe that all green apples are the definition for what a desirable red apple should be.

At the heart of this is something deeper than statistical ignorance, it's also the real risk that more people around you will start to more easily modify their own real common sense about how to make independent decisions. There would be pressure to statistically conform your ideas, and in a way that endangers your future. We have also seen this with the 2020 vaccine development, where the standards of success are rapidly disguised and changed (retroactively) in order to fit the results of spurious non-random trials. Or a-priori statistical tests or measures continuously mocked, and not by people with wide knowledge. We've seen lock-down orders and "scientific" criteria for easing them suddenly vanished once the medical community (who initially thought this virus was just the flu) decided that Black Lives Matters protests were more important. Medical professionals should not be in the businesses of staking their reputation on impulsively conflated virology science with their own political rulers.

Recall these conceptual rulers multi-dimensionally leaning on one another at times, too. A couple of years ago Johns Hopkins ranked the US as the most prepared for a pandemic. No one doubted it, but once an actual pandemic occurred, they instantly retracted the ranking in early 2020. Or Trump who praised Xi on his handling of the coronavirus as a boon for other policy objectives Trump wanted to accomplish with him. Xi then praised WHO for their public comments and response to the virus. And WHO then praised Trump. Everyone keen to praise one another initially, yet you were being subject to the most believable false ruler meant to benefit everyone but you.

Also it reinforces the wrong things that are, and the wrong things that are not. Such as when an unaccomplished influencer uses their status to endorse something they know little about. When this happens, and it oft-times does, what do you think will be the likely result?

Now let's look our sixth and final vignette of this book. There are often much larger (and unseen) forces looming, versus whatever rigid statistical model you are focused on: Statisticians have developed fine forecasting tools, which could work well in narrow set-ups (also given tight prerequisites for their use). You can apply cluster or variance adjustments automatically, or fine-tune seasonality, etc.

But these techniques again may not be everlast if you are simply looking at the problem entirely wrong: You may not need a scalpel, but a jack-hammer! You may need to consider the popular frameworks could eventually completely breakdown in "unpredictable" ways, rendering them useless when the much larger forces eventually arrive: For example, the longest economic recovery "in history" stopped short by a sudden "unforeseen" pandemic: And now we must wonder if the next pandemic will be 100 years from now, 50 years from now, next year, etc.

How will you ever again trust "seasonally adjusted" economic calculations, after the enormity of errors we saw in them, during 2020? And how will experts ever be trusted to forecast long-term disease outbreaks, when they were clearly unable to see a pandemic unfold real-time, before the world's eyes in 2020?

So our idea here is that we might not even know the basic direction of things, let alone need to worry about any specific false target: And generally in life, how do you know whether you are on the right path, or simply the most convenient path for you? What does it mean to discover the "love of your life" when in fact, you have yet to discover who you are?

How do you know you weren't swapped at birth, or once given someone else's medical lab results with your name on it? How do you know if you are not in someone else's dream right now... as you are reading this book?

You may never know, which is what makes our belief in being able to model many benign things (close to our destiny), even more precarious. But also exhilarating. Long spans of time have a way of deepen a randomly false understanding of who you are. You might question whether you are on the right path, where no GPS (a.k.a. other people's advice masquerading as the truth) exists.

Is there even one right path, or are there other drastically different paths of which you were unaware. A hamster thinking running towards the strawberry is actually getting ahead, instead of being controlled by a wheel. But in your case, the wheel is imagined, and therefore should be freely unimagined.

Only you know whether you can are making the best of the models and randomness you experience daily. Only you are trapped in your body and with your own decisions. You can internalize what makes for the best center of gravity in your own orbit. Travel the world, meet people. Having gone to over 50 countries myself, there is something from others that a library of data could never replace!

No one else is the same. Take for example the differences between two siblings, one graduating from college in 2007 versus another graduating from college in 2009. Or one passing away a year after 9/11 and the other passing away a year before 9/11. In both cases, just two years apart, yet their impressions or assessment of global risk are drastically different.

In this book we explored six broad and interlinked statistics themes: missing data, proximate bias, perception wedges, busted rulers, compounded errors, and simulators. Each one may play a role (some more; some less) when thinking through statistical problems in your own journey. Statistics is more than an argument over the best formula, and instead it has reduced itself to an argument to poorly influence thought.

So certainly learn the formulae to inform you, but not for the most obvious pedagogical reason. Learn them so that you can be more cautious about others who use the same. Use them, by learning when you need to down-play them.

Make sure to also not getting stymied with too many mathematical or logical processes that you don't make concrete progress yourself. There is just as much risk from indecision and confusion sometimes, as there is from decision and progress. Aim each day to replace some friends who believe they know you and want you to maintain your current guidance system, with new friends who are unique. Life, similar to how we choose to swig incoming news data, is a continuous work in progress.

We hope you enjoyed the varied eclectic illustrations to add meditation and enjoyment the underlying concepts. There is, after all, a deeper art to appreciating math.